SCIENTIFIC AMERICAN EDUCATIONAL PUBLISHING

SCIENTIFIC AMERICAN INVESTIGATES LIFE CYCLES

THE STRANGE LIFE CYCLE OF A KANGAROO

ANNA MCDOUGAL

Published in 2025 by The Rosen Publishing Group
in association with Scientific American Educational Publishing
2544 Clinton Street, Buffalo NY 14224

Library of Congress Cataloging-in-Publication

Names: McDougal, Anna, author. Title: The strange life cycle of a kangaroo / Anna McDougal.
Description: Buffalo, NY : Scientific American Educational Publishing,
[2025] | Series: Scientific American investigates life cycles |
Includes bibliographical references and index.
Identifiers: LCCN 2024011175 (print) | LCCN 2024011176 (ebook) |
ISBN 9781725350885 (library binding) |
ISBN 9781725350878 (paperback) | ISBN 9781725350892 (ebook)
Subjects: LCSH: Kangaroos–Juvenile literature. | Kangaroos–Life cycles–Juvenile literature. |
Life cycles (Biology)–Juvenile literature.
Classification: LCC QL737.M35 M344 2025 (print) | LCC QL737.M35 (ebook) |
DDC 599.2/22156–dc23/eng/20240320
LC record available at https://lccn.loc.gov/2024011175
LC ebook record available at https://lccn.loc.gov/2024011176

Portions of this work were originally authored by Barbara M. Linde and published as *The Bizarre Life Cycle of a Kangaroo.*
All new material in this edition authored by Anna McDougal.

Designer: Tanya Dellaccio Keeney
Editor: Caitie McAneney

Photo credits: Cover Karen Brough/Shutterstock.com; p. 5 Rafael Ramirez Lee/Shutterstock.com; p. 7 (top) ariff gizan/Shutterstock.com; p. 7 (bottom) Anan Kaewkhammul/Shutterstock.com; p. 9 Nomadic Exposure/Shutterstock.com; p. 11 Auscape International Pty Ltd/Alamy.com; p. 13 K.A.Willis/Shutterstock.com; p. 15 Kjuuurs/Shutterstock.com; p. 17 Karen Brough/Shutterstock.com; p. 19 kyslynskahal/Shutterstock.com; p. 20 Agami Photo Agency/Shutterstock.com.

Printed in the United States of America

CPSIA compliance information: Batch #CSSA25. For Further Information contact Rosen Publishing at 1-800-237-9932.

CONTENTS

'ROOS RULE! 4
MEET THIS MARSUPIAL 6
KANGAROO MOBS 8
TINY BABIES. 10
SAFE WITH MOM 12
JUMPING JOEYS. 14
DOUBLE THE FUN 16
KANGAROO LIFE 18
THE CYCLE REPEATS. 20
GLOSSARY 22
FOR MORE INFORMATION 23
INDEX 24

Words in the glossary appear in **bold** type the first time they are used in the text.

'ROOS RULE!

What's that tall animal hopping across the Australian **Outback**? It's the amazing kangaroo. This **mammal** has a strong tail and powerful back legs. Kangaroos can leap up to 30 feet (9.1 m) away in just one jump. That's far! You can also identify kangaroos by their two short front legs and the wide pouch on the bellies of females.

This pouch has an important purpose! It's part of the strange life cycle of the kangaroo. Read on to find out more about the 'roo!

A kangaroo's tail helps it balance and move.

FUN FACT

ADULT KANGAROOS CAN BE 6 FEET (1.8 M) TALL—AS TALL AS AN ADULT MAN!

MEET THIS MARSUPIAL

A kangaroo is more than just a mammal. It's a kind of mammal called a marsupial. A female marsupial has a pouch made of skin and fur on her stomach. Instead of growing inside their mother's body, baby marsupials do most of their growing inside this pouch.

Kangaroos have other unusual features, too. They can't walk on all four legs or move backward easily. Kangaroos can't move their legs separately while on land—only when they swim. They are great at hopping though!

FUN FACT

A BABY KANGAROO IS CALLED A JOEY.

A kangaroo's powerful back legs can take it far. It can leap more than 6 feet (1.8 m) high in one bound!

KANGAROO MOBS

A group of kangaroos is called a mob. Kangaroo mobs may have 100 or more members! They include adult males, adult females, and joeys. An adult male kangaroo is called a buck or boomer. A female is a doe or flyer.

The adult males often fight with each other. The largest, strongest male wins and becomes the leader. This male can **mate** with more than one female each season. Kangaroo females can have babies at any time of the year.

Kangaroos are social animals, or those that like being around others.

FUN FACT

KANGAROOS WILL FIGHT IF THEY SENSE A THREAT. THEY USE THEIR STRONG FEET TO POUND THE GROUND. THEY WILL ALSO KICK OR BITE.

TINY BABIES

When a female kangaroo is about 18 months old, she can start having babies. She has one at a time. After about 34 days, a jelly-bean-sized baby is born. However, it's not ready for the world yet.

The newborn is blind and doesn't have any fur. Its brain, heart, lungs, and other organs aren't fully **developed** yet. Its back legs aren't, either. The tiny baby is completely **dependent** on its mother and will spend the next few months inside her pouch.

This tiny newborn kangaroo is growing in its mother's pouch.

FUN FACT

A BABY KANGAROO IS ABOUT 1 INCH (2.5 CM) LONG AND WEIGHS ABOUT 0.03 OUNCE (1 GRAM).

SAFE WITH MOM

How does the tiny baby get to its mother's pouch? It crawls up its mother's furry stomach! Some scientists think it uses a strong sense of smell to find the way. The baby uses its forelegs (front legs) to climb. The baby stays in the pouch, drinking milk its mother makes. Its fur and body parts grow.

After about 7 to 9 months, the joey starts to come out of the pouch for short periods of time. It's time to explore the world!

Joeys and their mothers are inseparable for the first few months of life.

FUN FACT

NEWBORN KANGAROOS NEED THE MUSCLES IN THEIR MOTHER'S POUCH TO PUMP MILK INTO THEIR THROATS. THEY CAN'T SUCK OR SWALLOW YET.

JUMPING JOEYS

Joeys that leave their pouch don't stray very far from their mothers. The watchful mother is only a few hops away. The joey may hop around and eat grass. Then, it returns to the pouch to sleep and drink milk. Other mothers and joeys are nearby.

When the joey is around 11 months old, it leaves the pouch for good. Now it's called a joey-at-foot. It will still be about another year for the joey to fully stray from its mother.

If people or other animals get too close, the mothers **defend** the joeys.

FUN FACT

MOTHER KANGAROOS COMMUNICATE WITH THEIR YOUNG THROUGH CLUCKS AND CLICKS.

DOUBLE THE FUN

A mother kangaroo can care for two babies at once. The joey may leave the pouch, but it still needs its mother for milk. The mother may have another baby at this time. This newborn crawls into its mother's pouch to grow, even though the older joey still needs her.

A mother kangaroo cares for only two joeys at a time. However, she may have another **embryo** waiting inside her. The embryo won't develop until the older joey is gone for good.

Kangaroo mothers and joeys are as close as can be!

FUN FACT

A FEMALE KANGAROO'S BODY MAKES A DIFFERENT TYPE OF MILK FOR EACH JOEY, THE NEWBORN AND THE OLDER ONE!

KANGAROO LIFE

All kangaroos are native only to Australia and its islands. The three most common **species** of kangaroos are the eastern gray, the western gray, and the red. They may live in small or large mobs, or sometimes alone.

Kangaroos are **herbivores**. They spend their days grazing, or eating, the grass from fields or leaves from trees. The mob travels together to find food. Kangaroos eat in the cool mornings and evenings, and rest during the hot afternoons. They may swim across a stream to find food or escape an enemy.

The red kangaroo is the largest species. You can identify the males by their orange-red coloring.

FUN FACT

KANGAROOS CAN GO FOR MONTHS WITHOUT WATER. THAT'S A USEFUL **ADAPTATION** FOR THE DRY AUSTRALIAN OUTBACK!

THE CYCLE REPEATS

Once a female joey grows a bit more, she's ready to have her own babies. The cycle begins again! Kangaroos aren't the only mammals to have this crazy life cycle. Wallabies are like cousins to kangaroos. They also have tiny babies that grow in their pouches!

Living with mom for so long has its perks. Joeys have a constant source of milk. They also have protection from outside threats. That's life in the Outback!

Follow the life cycle of a kangaroo from embryo to adult!

LIFE CYCLE OF A KANGAROO

GLOSSARY

adaptation: A change in a type of animal that makes it better able to live in its surroundings.

communicate: To share thoughts or feelings by sound, movement, or writing.

defend: To guard from harm.

dependent: To need the help of another to live.

develop: To grow and change.

embryo: An animal in the early stages of growth, usually before it's born.

herbivore: An animal that eats only plants.

mammal: A warm-blooded animal that has a backbone and hair, breathes air, and feeds milk to its young.

mate: To come together to make babies.

Outback: A remote, rural area in Australia or New Zealand.

species: A group of living things that are all of the same kind.

threat: Something likely to cause harm.

FOR MORE INFORMATION

Books

Holmes, Kirsty. *Life Cycle of a Kangaroo.* New York, NY: KidHaven Publishing, 2022.

Nilsen, Genevieve. *Kangaroo Joeys.* Minneapolis, MN: Jump! Inc., 2022.

Websites

Kangaroo
kids.nationalgeographic.com/animals/mammals/facts/kangaroo
Check out fun facts about kangaroos from National Geographic Kids.

Marsupial
kids.britannica.com/kids/article/marsupial/353437
Learn more about marsupial babies.

INDEX

adult female (doe, flyer), 8, 10, 11, 12, 13, 14, 15, 16, 17, 20, 21

adult male (buck, boomer), 8

diet, 16, 19, 20

fighting, 9

habitat, 18

joey, 7, 10, 11, 12, 13, 14, 15, 16, 17, 20

joey-at-foot, 14, 21

jumping, 4, 7

legs, 4, 7

marsupial, 6

mob, 8, 18

mother's milk, 12, 13, 14, 16, 17, 21

movement, 6, 7

physical description, 4, 5

pouch, 4, 6, 10, 11, 12, 13, 14, 16

tail, 4, 5

wallaby, 20